www.oxfordowl.co.uk
AF470580
Well done!
Good work!
Busy worker!
Smart work!
Superstar!
Brilliant!
I did really well!
Well done!
Amazing!
Star!
Good work!
CE
WARNING!
NOT SUITABLE FOR
CHILDREN UNDER
36 MONTHS DUE
TO SMALL PARTS
WHICH PRESENT A
CHOKING HAZARD.

Jenny Ackland

At Home With COUNTING

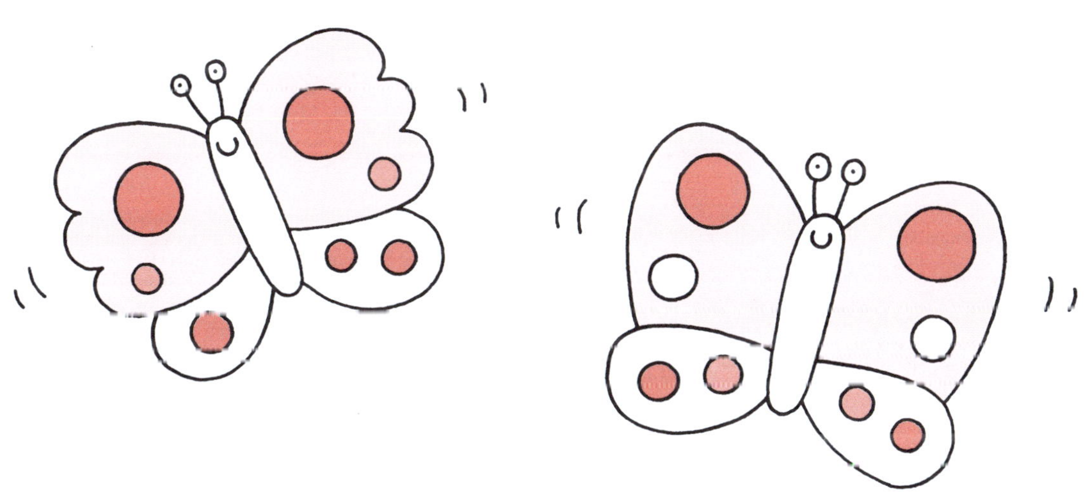

OXFORD
UNIVERSITY PRESS

Introduction

The *At Home With* workbooks introduce and reinforce key numeracy and literacy concepts for primary school children. They provide lots of opportunities to develop the key skills that are the basis of primary school curriculum work. The workbooks are available in three levels: 3–5 years, 5–7 years, and 7–9 years. The activities are fun and are designed to stimulate discussion, as well as practical skills. Some children will be able to complete the activities alone, after initial discussion; others may benefit from adult support throughout. All children will enjoy rewarding themselves with a sticker when they reach the end of an activity.

Using the book

At Home With Counting offers a variety of activities which focus, in particular, on the following:

- revision of numbers 1 to 5

- counting and recognizing numbers 0 to 10

- sequencing numbers up to 10

- understanding 'more' and 'less'

- basic addition and subtraction

- recognition of numbers 1 to 20 (where appropriate).

Oxford University Press
Great Clarendon Street, Oxford OX2 6DP

Oxford University Press is a department of the University of Oxford.
It furthers the University's objective of excellence in research, scholarship, and education by publishing worldwide in
Oxford New York Auckland Cape Town Dar es Salaam Hong Kong Karachi
Kuala Lumpur Madrid Melbourne Mexico City Nairobi New Delhi
Shanghai Taipei Toronto

With offices in
Argentina Austria Brazil Chile Czech Republic France Greece Guatemala
Hungary Italy Japan Poland Portugal Singapore South Korea Switzerland
Thailand Turkey Ukraine Vietnam

Oxford is a registered trade mark of © Oxford University Press
in the UK and in certain other countries

© Jenny Ackland 2005
The moral rights of the author have been asserted
Database right Oxford University Press (maker)
First published 2005
Reissued 2009
This edition 2012

British Library Cataloguing in Publication Data
Data available

ISBN: 978 0 19 273324 5

1 3 5 7 9 10 8 6 4 2

Designed by Red Face Design
Illustrations by Mark Brierley, Sue Cony
Printed in China

Contents

Numbers 1 to 5

Write the numbers and count the animals.

1 1 1 1 1

2 2 2 2 2

3 3 3 3 3

4 4 4 4 4

5 5 5 5 5

Sets of 5

Circle and colour the sets of 5.

Missing numbers

Fill in the missing numbers.

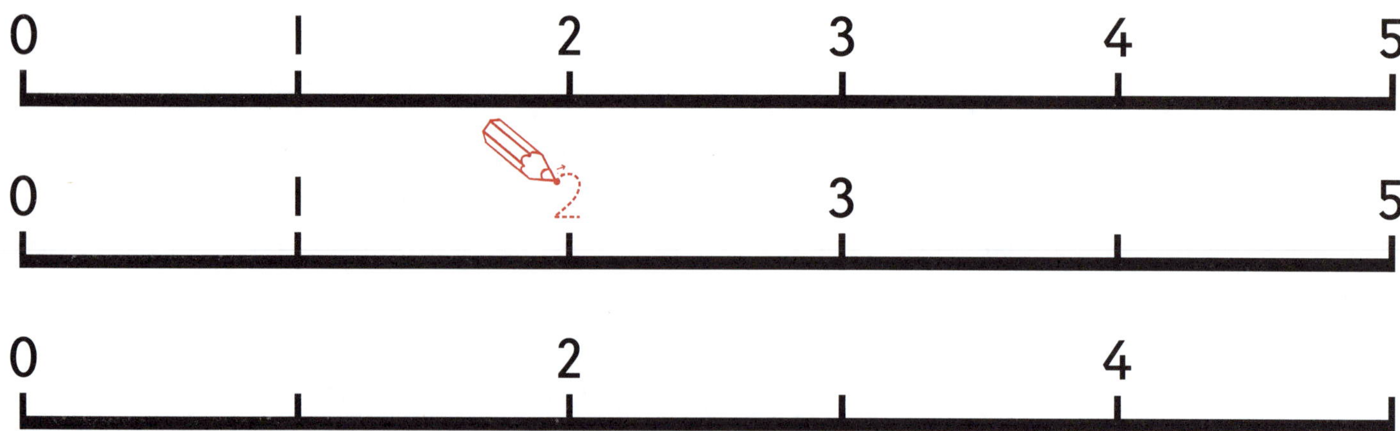

Read the numbers and draw the spots.

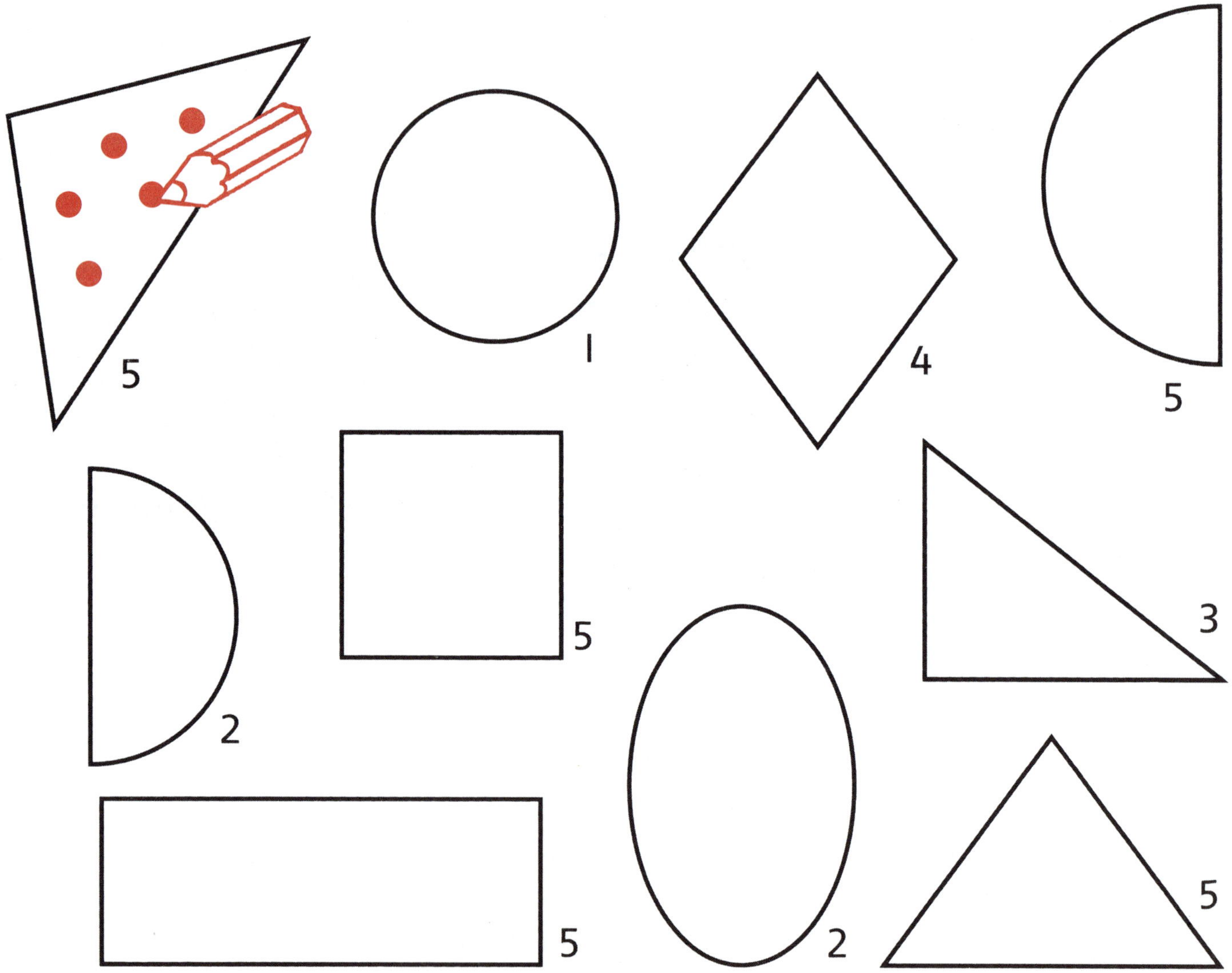

Recognizing sets

Read each sentence and fill in the missing number.

There are _________ cars.

There are _________ birds.

There are _________ dogs.

There are _________ trees.

Number 6

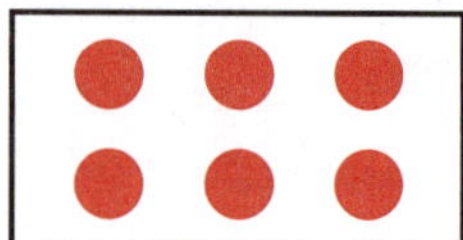

Circle and colour the sets of 6.

Counting 6

1 1 1 1 1

2 2 2 2 2

3 3 3 3 3

4 4 4 4 4

5 5 5 5 5

6 6 6 6 6

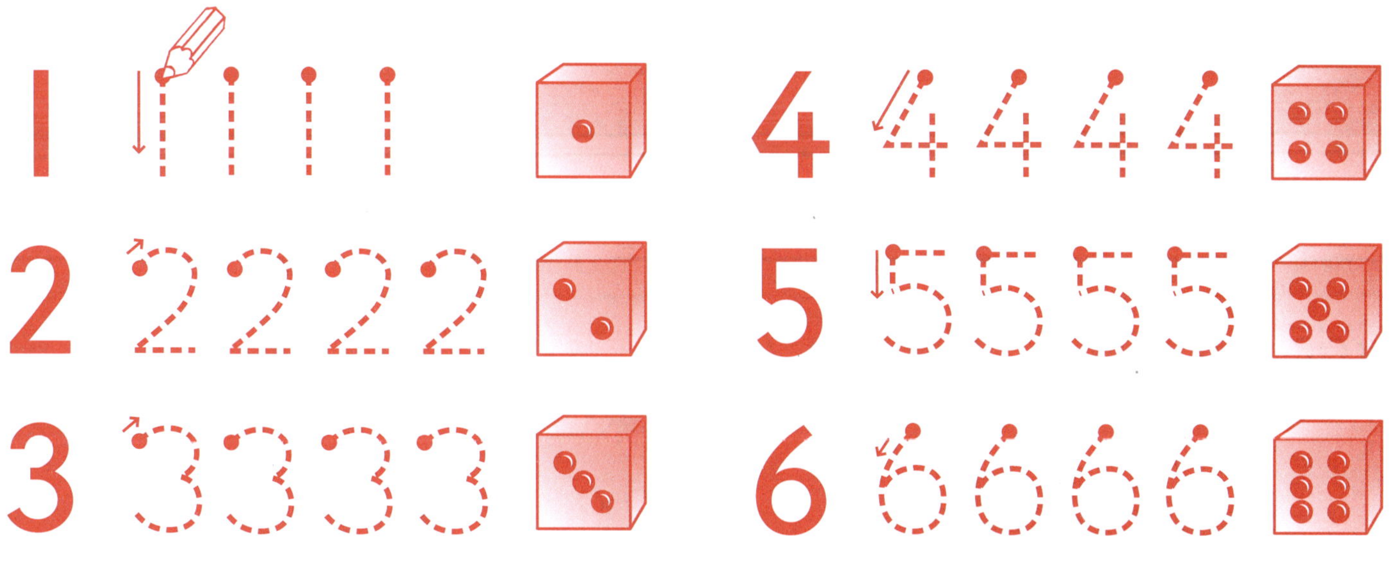

Give each cat 6 whiskers.

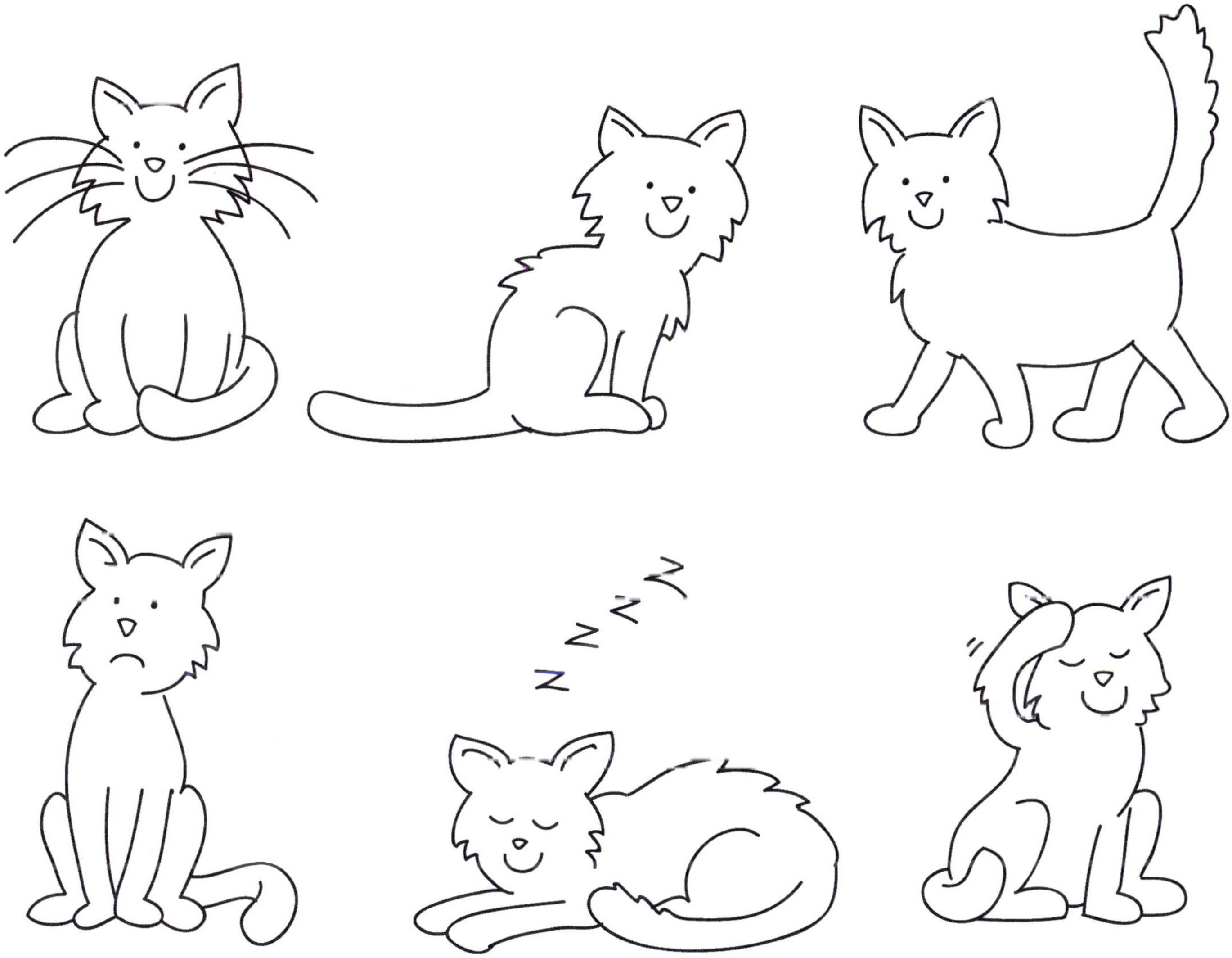

Number 7

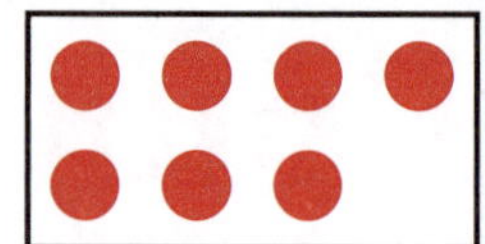

Circle and colour the sets of 7.

Counting 7

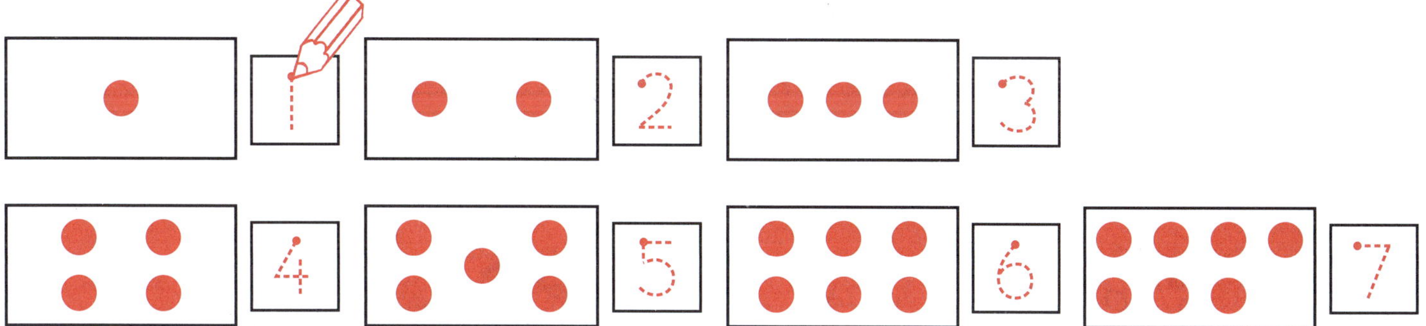

Draw and colour 7 balloons.

Sets of 7

Circle and colour the sets of 7.

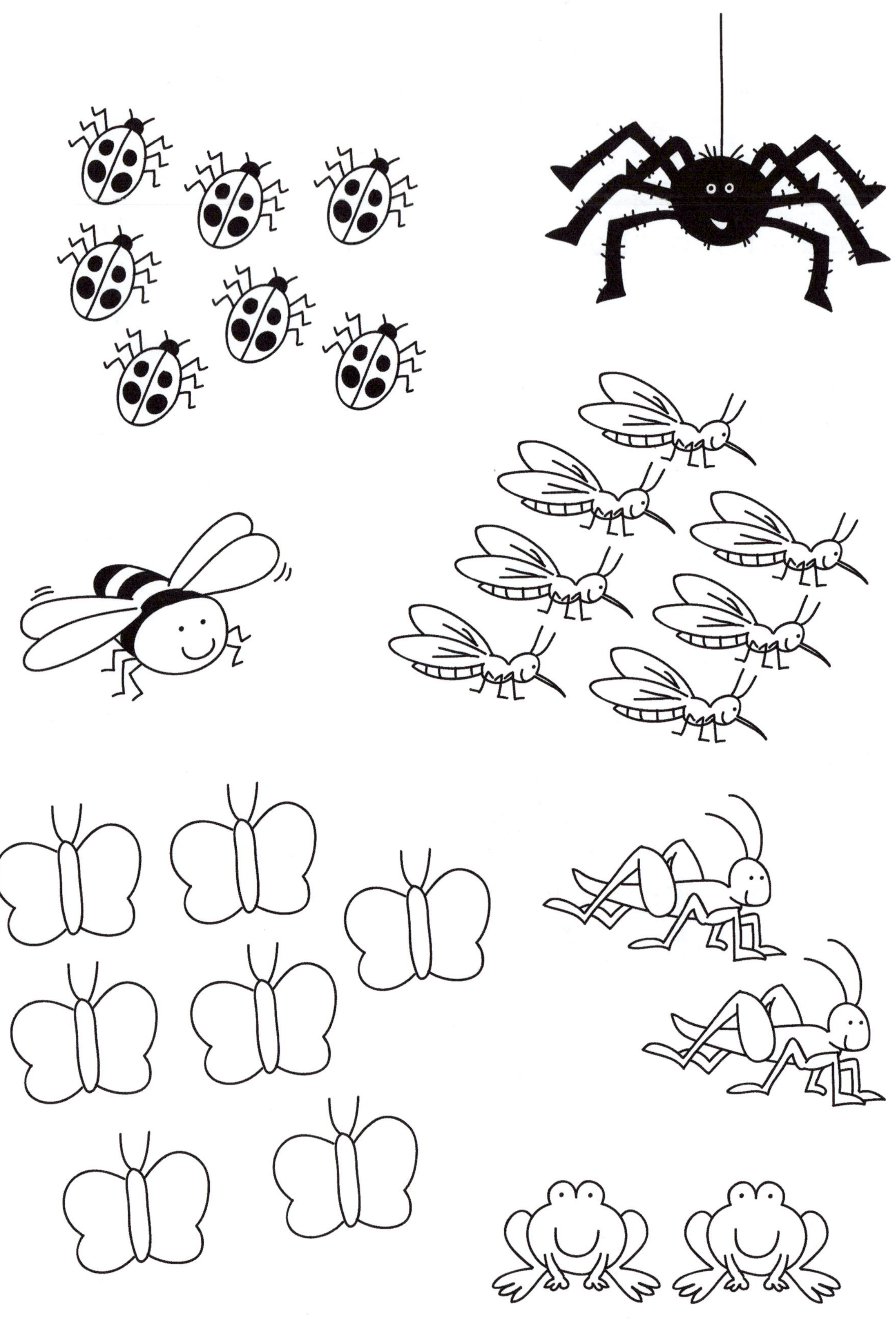

Counting to 7

How many birds are there? Circle the right number.

Number 8

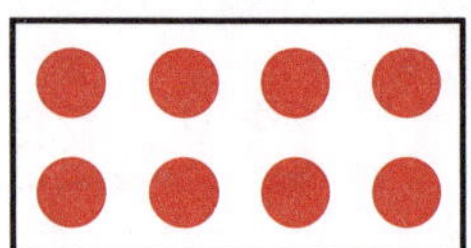

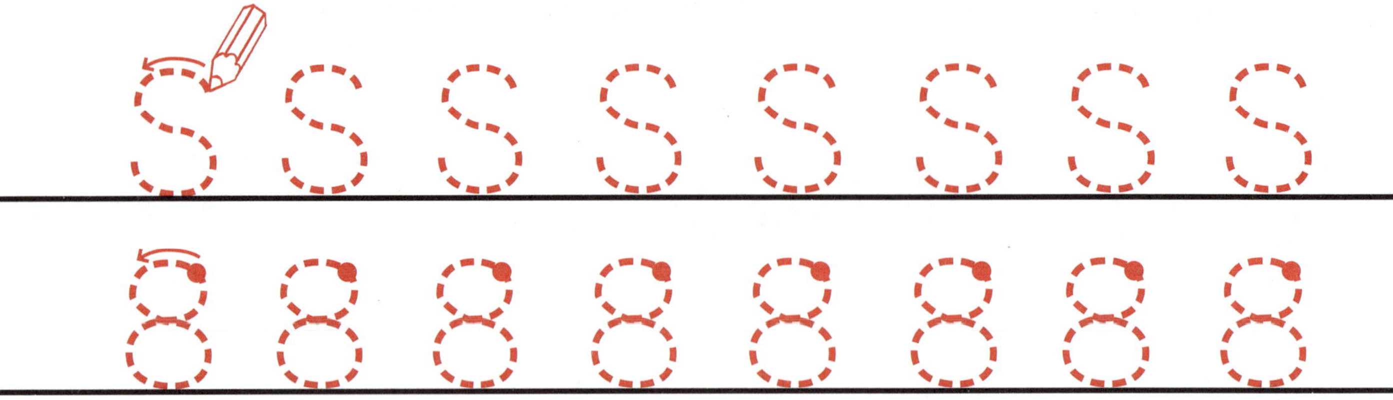

Give this bug 8 legs. Colour them.

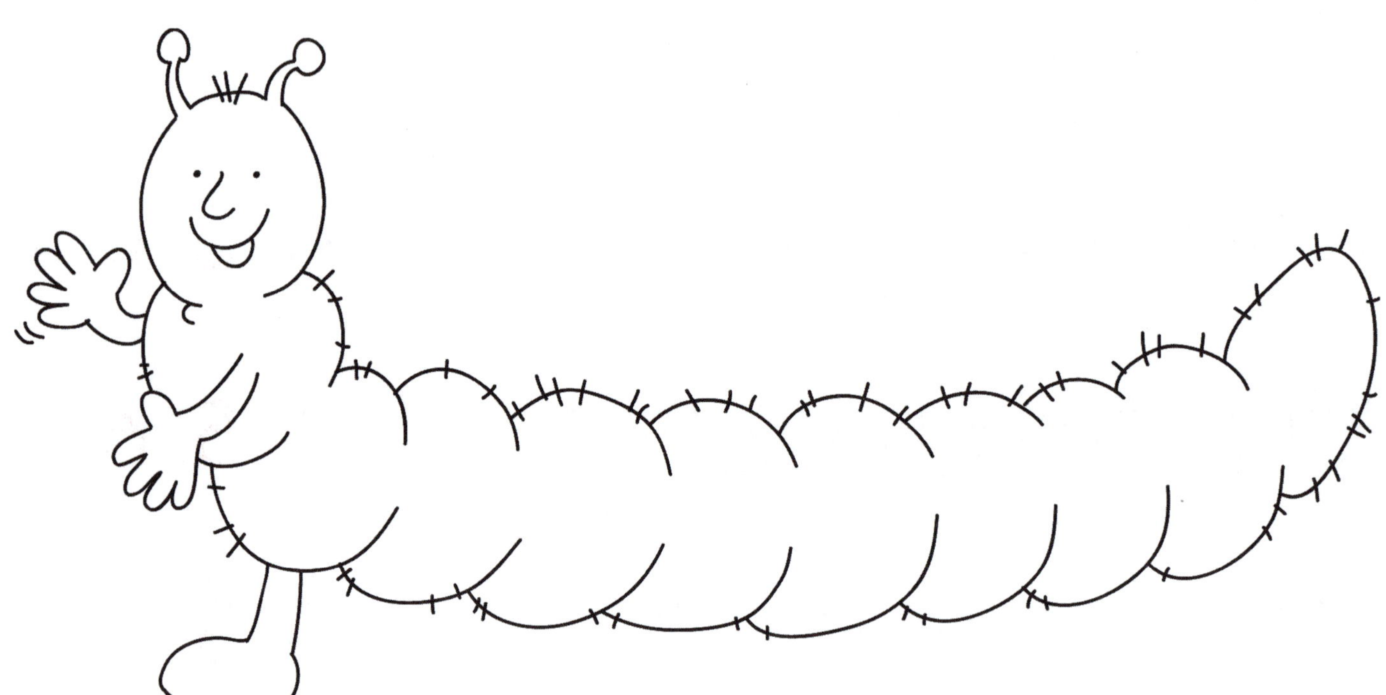

Counting to 8

Draw the dots on these dominoes.

1 2 3 4

5 6 7 8

Count the people and circle the right number.

1 2 ③ 4 5 6 7 8

1 2 3 4 5 6 7 8

1 2 3 4 5 6 7 8

1 2 3 4 5 6 7 8

1 2 3 4 5 6 7 8

Number 9

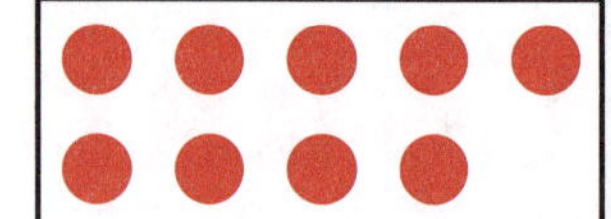

     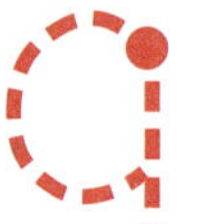 

This girl is counting worms. Draw 9 worms.

Counting to 9

Colour 6 socks.

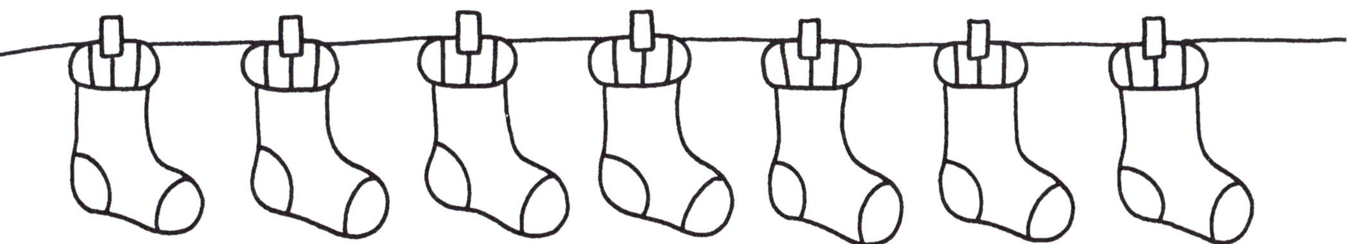

Colour 7 T-shirts.

Colour 8 gloves.

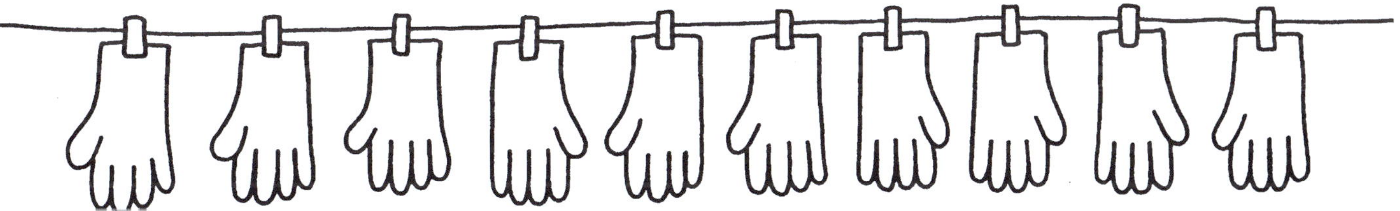

Colour 9 teddies.

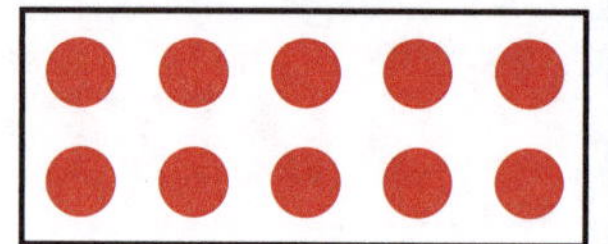

Number 10

Draw 10 candles. Colour them.

Counting to 10

Count the balloons by joining them to the number line.
Then colour them in.

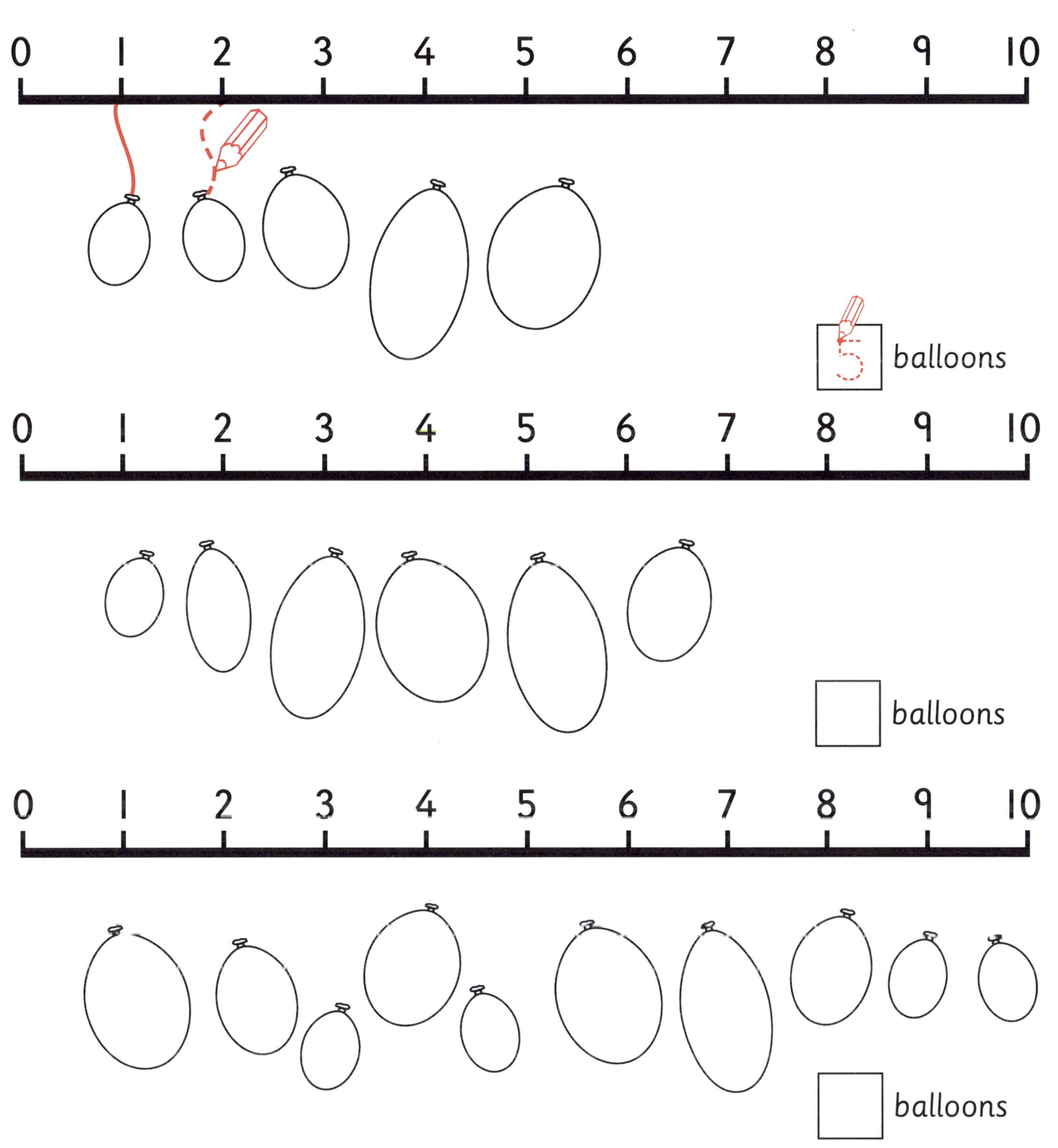

Sequencing 0 to 10

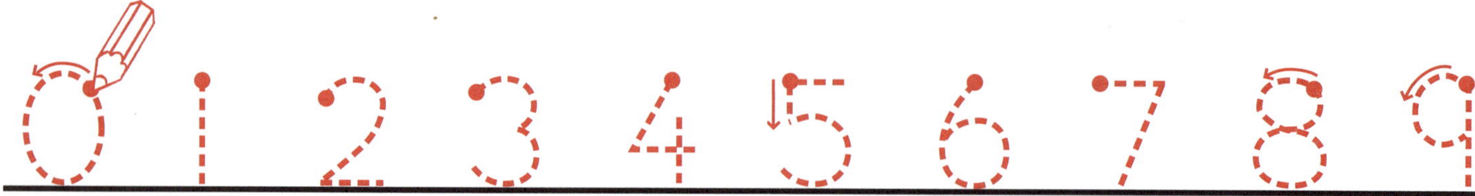

Join the dots in number order.

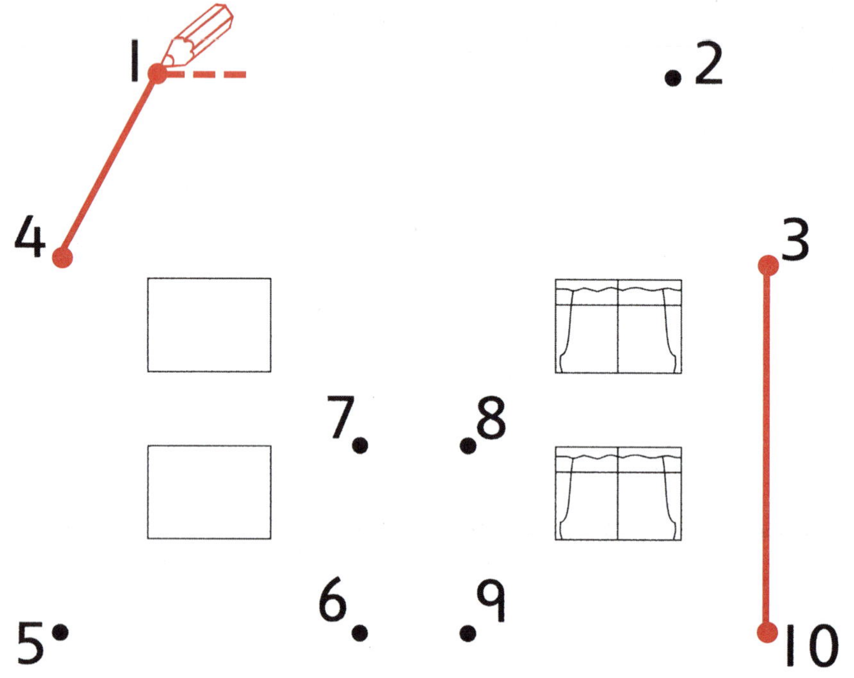

Join the dots in number order.

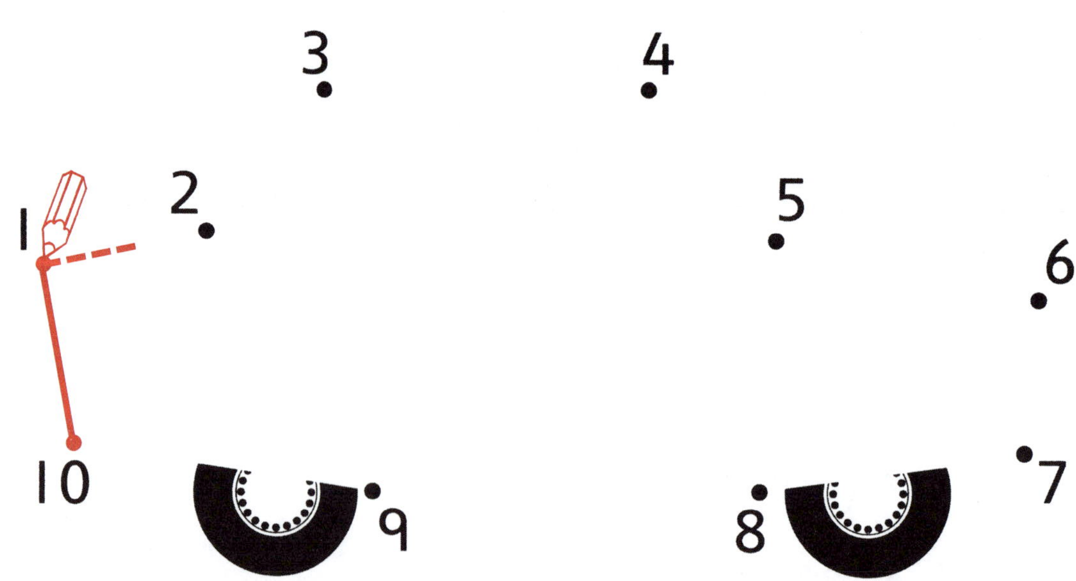

Counting rhyme

Say the rhyme. Point to the numbers as you say them.

1   2   3   4   5

One, two, three, four, five,

Once I caught a fish alive,

6   7   8   9   10

Six, seven, eight, nine, ten,

Then I let it go again.

Join the dots in number order.

Counting

0 1 2 3 4 5 6 7 8 9 10

Number the stones and count how many jumps.

Counting on

Add one more jump. Write the final number.

Add two more jumps. Write the final number.

Adding sets

Colour 5 legs green.

Colour 1 leg blue.

Altogether there are ☐ legs.

Colour 8 legs red.

Colour 2 legs yellow.

Altogether there are ☐ legs.

Colour 7 legs orange.

Colour 2 legs brown.

Altogether there are ☐ legs.

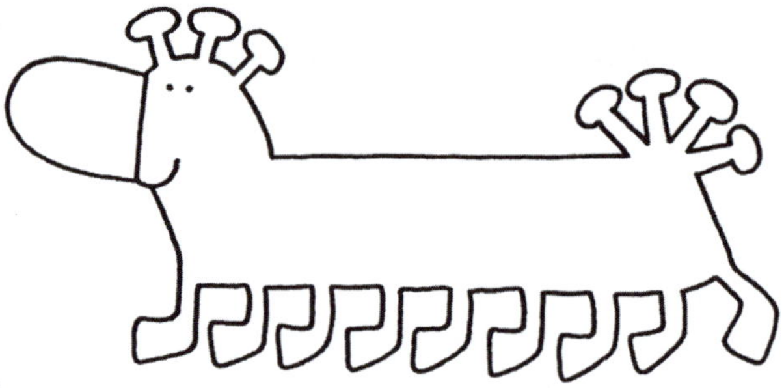

Colour 6 legs pink.

Colour 4 legs black.

Altogether there are ☐ legs.

Number bonds

Colour 9 spots red.

Colour 1 spot blue.

Altogether there are ☐ spots.

Colour 8 spots green.

Colour 2 spots red.

Altogether there are ☐ spots.

Colour 7 spots pink.

Colour 3 spots yellow.

Altogether there are ☐ spots.

Colour 6 spots orange.

Colour 4 spots purple.

Altogether there are ☐ spots.

Add 1 more

0   1   2   3   4   5   6   7   8   9   10

Add one more biscuit to each plate.

Now there are ⟦4⟧ biscuits.

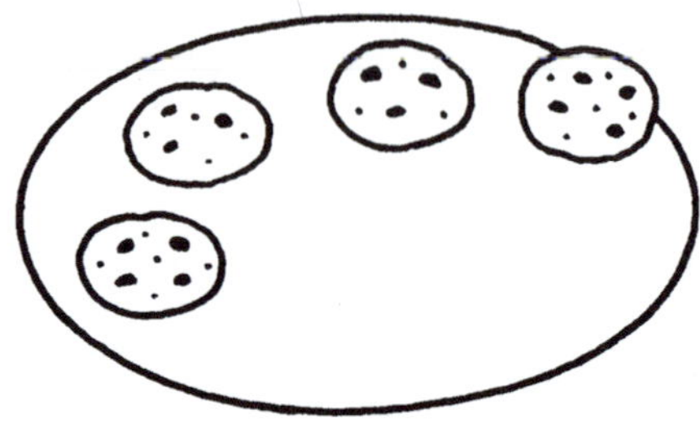

Now there are ⟦ ⟧ biscuits.

Now there are ⟦ ⟧ biscuits.

Now there are ⟦ ⟧ biscuits.

Now there are ⟦ ⟧ biscuits.

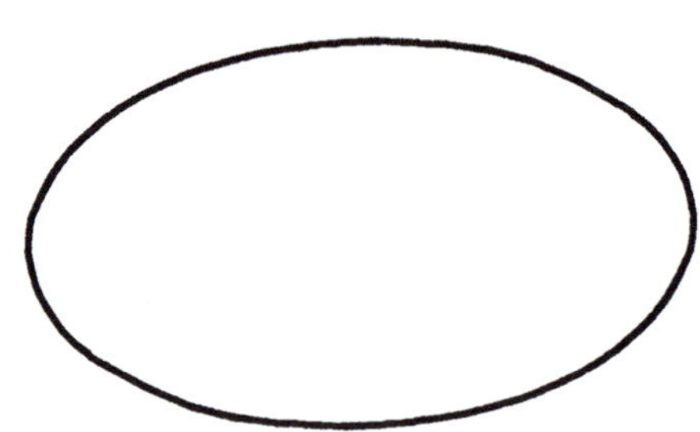

Now there is ⟦ ⟧ biscuit.

More than

Fill in the missing numbers.

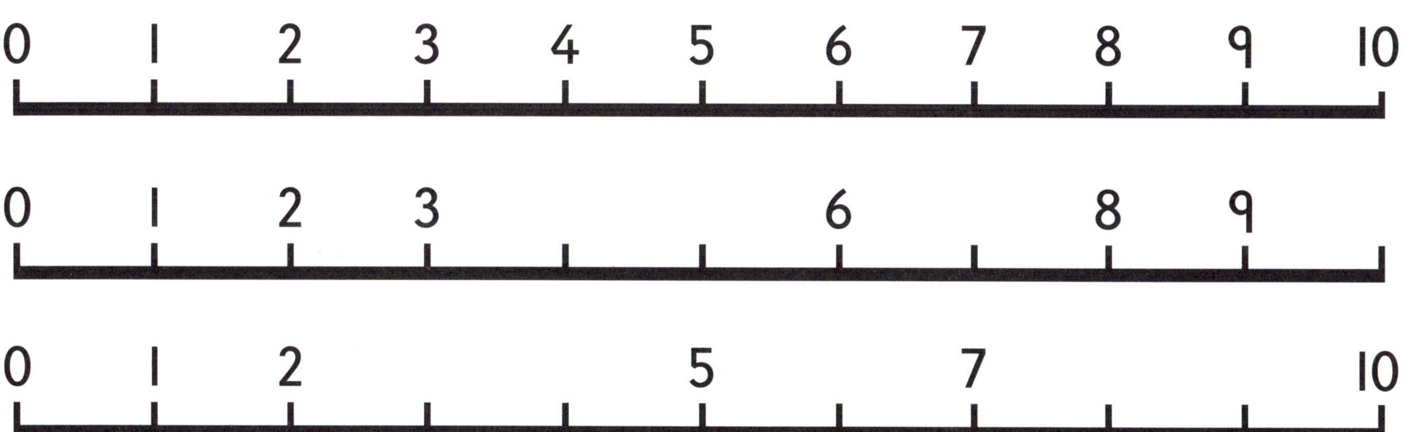

Draw the beads, then write the answer.

One more than 6 is [7]

One more than 5 is [ ]

One more than 7 is [ ]

One more than 4 is [ ]

Two more than 6 is [ ]

Two more than 5 is [ ]

Two more than 7 is [ ]

Two more than 4 is [ ]

Less than

Fill in the missing numbers.

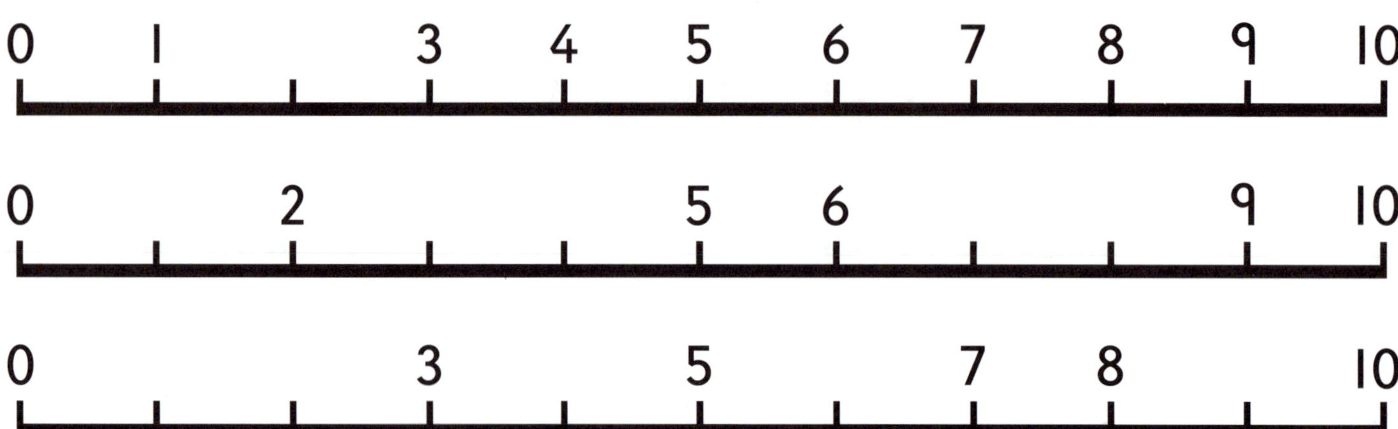

Draw the beads, then write the answer.

One less than 6 is ⬜ 5

One less than 5 is ⬜

One less than 7 is ⬜

One less than 4 is ⬜

Two less than 6 is ⬜

Two less than 5 is ⬜

Two less than 7 is ⬜

Two less than 4 is ⬜

Take 2 away

0 1 2 3 4 5 6 7 8 9 10

Take 2 pieces of fruit away from each plate.
Count how many are left and fill in the number.

Now there are ⬚ bananas. Now there are ⬚ apples.

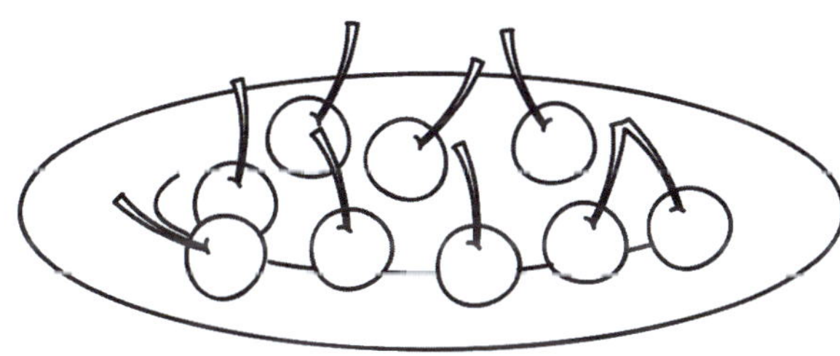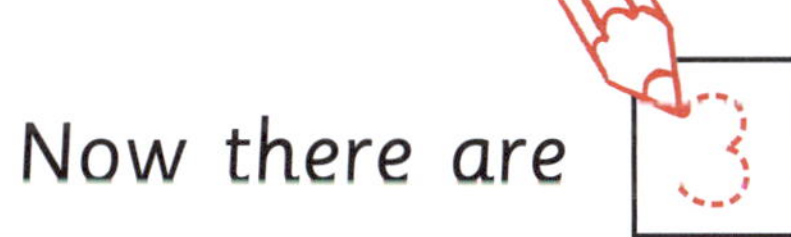

Now there are ⬚ cherries. Now there are ⬚ grapes.

Now there are ⬚ oranges. Now there is ⬚ pear.

Numbers 0 to 20

Fill in the missing numbers.

0 1 2 3 4 5 6 7 8 9 10 11 12 13 14 15 16 17 18 19 20

0   3   6   10   12   17

Number the racing cars in order.

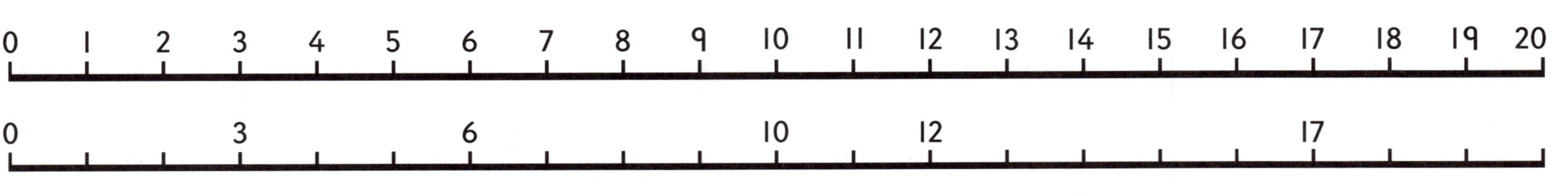

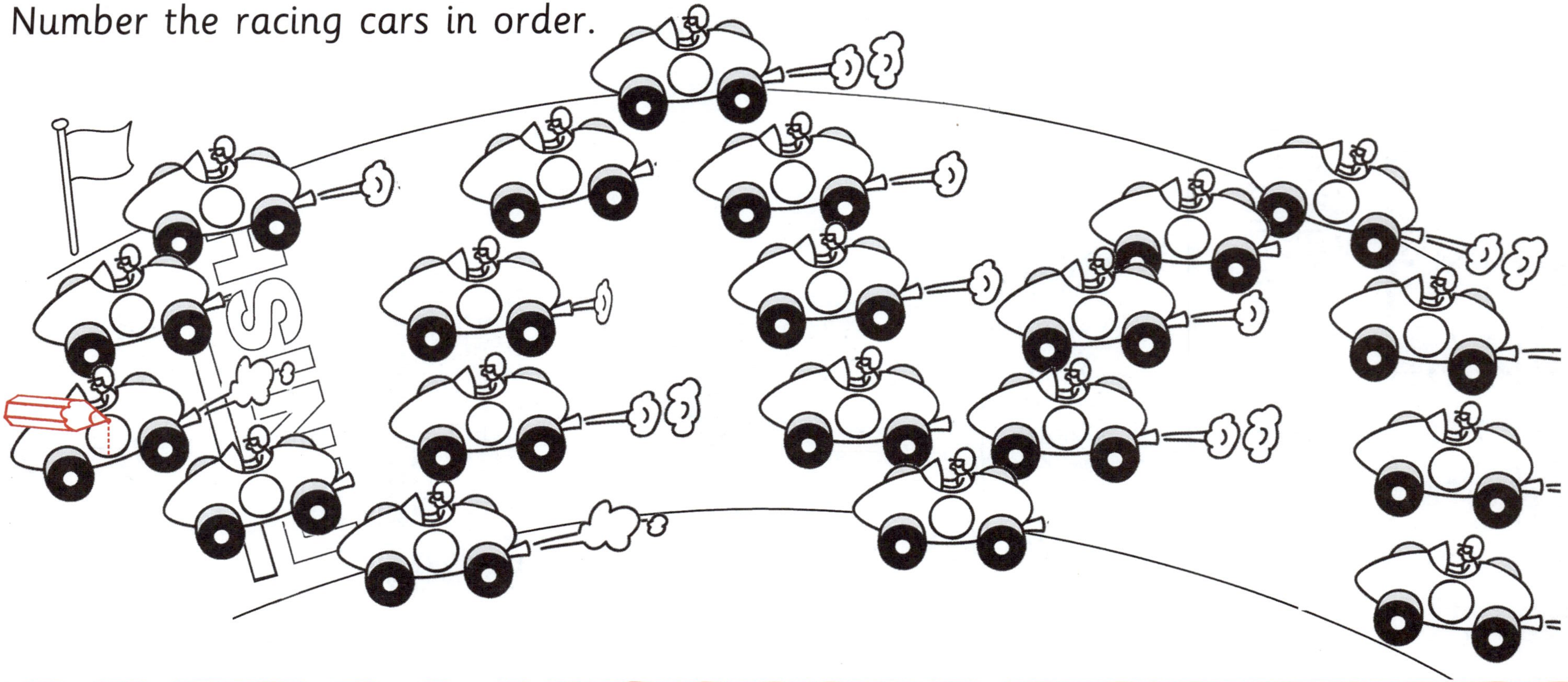

2 at a time

Climb the steps two at a time.
Write the number on top of the step.

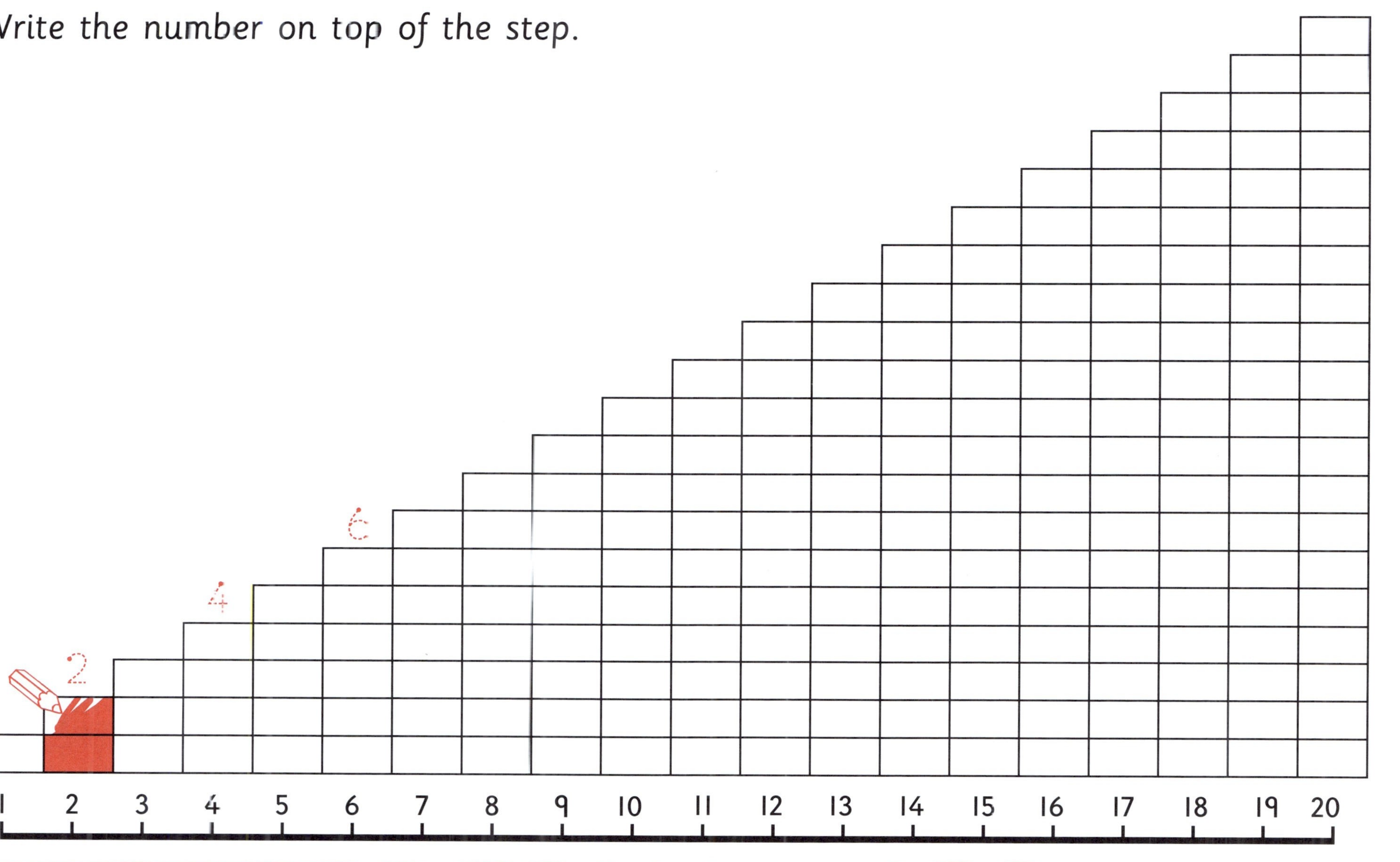

Summary of skills

Title	Page	Summary
Numbers 1 to 5	4	Revising numbers 1 to 5
Sets of 5	5	Recognizing sets of 5
Missing numbers	6	Completing number lines, counting spots
Recognizing sets	7	Finding sets and completing sentences
Number 6	8	Recognizing sets of 6
Counting 6	9	Counting and understanding 6
Number 7	10	Recognizing sets of 7
Counting 7	11	Writing numerals up to 7, drawing a set of 7
Sets of 7	12	Recognizing sets of 7
Counting to 7	13	Circling the right number
Number 8	14	Writing and understanding 8
Counting to 8	15	Counting and writing up to 8
Number 9	16	Writing and understanding 9
Counting to 9	17	Colouring objects corresponding to number
Number 10	18	Writing and understanding 10
Counting to 10	19	Joining balloons to number lines
Sequencing 0 to 10	20	Dot to dot
Counting rhyme	21	Counting and pointing to numbers, dot to dot
Counting	22	Counting by jumping from stone to stone
Counting on	23	Counting on by jumping one or two jumps
Adding sets	24	Finding the sum of two sets
Number bonds	25	Adding number bonds to 10
Add 1 more	26	Practical demonstration of adding 1 more
More than	27	Understanding the term 'more than'
Less than	28	Understanding the term 'less than'
Take 2 away	29	Practical demonstration of taking 2 away
Numbers 0 to 20	30	Sequencing numbers 1 to 20 (could lead to a discussion of ordinal numbers)
2 at a time	31	Practical demonstration of counting in 2s

At Home With 3–5 years

abc	Reading
Colours	Shape & Size
Counting	Sounds & Rhymes
Letter Forms	Pattern & Shape
Numbers	Writing

At Home With 5–7 years

English	Phonics
French	Spelling 1
Handwriting 1	Spelling 2
Handwriting 2	Times Tables
Maths	

At Home With 7–9 years

English	Punctuation
French	Reasoning Skills – Verbal
Grammar	Reasoning Skills – Non Verbal
Maths	Spanish
Mental Maths	